~A BINGO BOOK~

Weather Bingo Book

COMPLETE BINGO GAME IN A BOOK

Written By Rebecca Stark
Educational Books 'n' Bingo

WEATHER BINGO DIRECTIONS

INCLUDED:

List of Terms

Templates for Additional Terms and Clues

2 Clues per Term

30 Unique Bingo Cards

Markers

1. **Either cut apart the book or make copies of ALL the sheets. You might want to make an extra copy of the clue sheets to use for introduction and review. Keep the sheets in an envelope for easy reuse.**

2. Cut apart the call cards with terms and clues.

3. Pass out one bingo card per student. There are enough for a class of 30.

4. Pass out markers. You may cut apart the markers included in this book or use any other small items of your choice.

5. Decide whether or not you will require the entire card to be filled. Requiring the entire card to be filled provides a better review. However, if you have a short time to fill, you may prefer to have them do the just the border or some other format. Tell the class before you begin what is required.

6. There are 50 terms. Read the list before you begin. If there are any terms that have not been covered in class, you may want to read to the students the term and clues before you begin.

7. There is a blank space in the middle of each card. You can instruct the students to use it as a free space or you can write in answers to cover terms not included. Of course, in this case you would create your own clues. (Templates provided.)

8. Shuffle the cards and place them in a pile. Two or three clues are provided for each term. If you plan to play the game with the same group more than once, you might want to choose a different clue for each game. If not, you may choose to use more than one clue.

9. Be sure to keep the cards you have used for the present game in a separate pile. When a student calls, "Bingo," he or she will have to verify that the correct answers are on his or her card AND that the markers were placed in response to the proper questions. Pull out the cards that are on the student's card keeping them in the order they were used in the game. Read each clue as it was given and ask the student to identify the correct answer from his or her card.

10. If the student has the correct answers on the card AND has shown that they were marked in response to the *correct questions,* then that student is the winner and the game is over. If the student does not have the correct answers on the card OR he or she marked the answers in response to *the wrong questions,* then the game continues until there is a proper winner.

11. If you want to play again, reshuffle the cards and begin again.

Have fun!

TERMS

air mass	gale
air pressure	hail
blizzard	humidity
breeze	hurricane
calm	latitude
Celsius	lightning
cirrus	meteorologist
climate	overcast
cloud	precipitation
cold	rain
condensation	rainbow
cumulus	revolution
deserts	seasons
dew	snow
direction(s)	storm(s)
downpour	stratus
drizzle	sun
drought	Temperate
equator	temperature
evaporation	thunder
Fahrenheit	tornado
flurries	Tropics
fog	water cycle
freeze(s)	weather
front	wind

Additional Terms

Choose as many terms as you would like and write them in the squares. Repeat each as desired. Cut out the squares and randomly distribute them to the class. Instruct the students to place the square on the center space of the card.

Weather Bingo

© Barbara M. Peller

Clues for
Additional Terms

Write two or three clues for each new term.

_____ 1. 2. 3.	_____ 1. 2. 3.
_____ 1. 2. 3.	_____ 1. 2. 3.
_____ 1. 2. 3.	_____ 1. 2. 3.

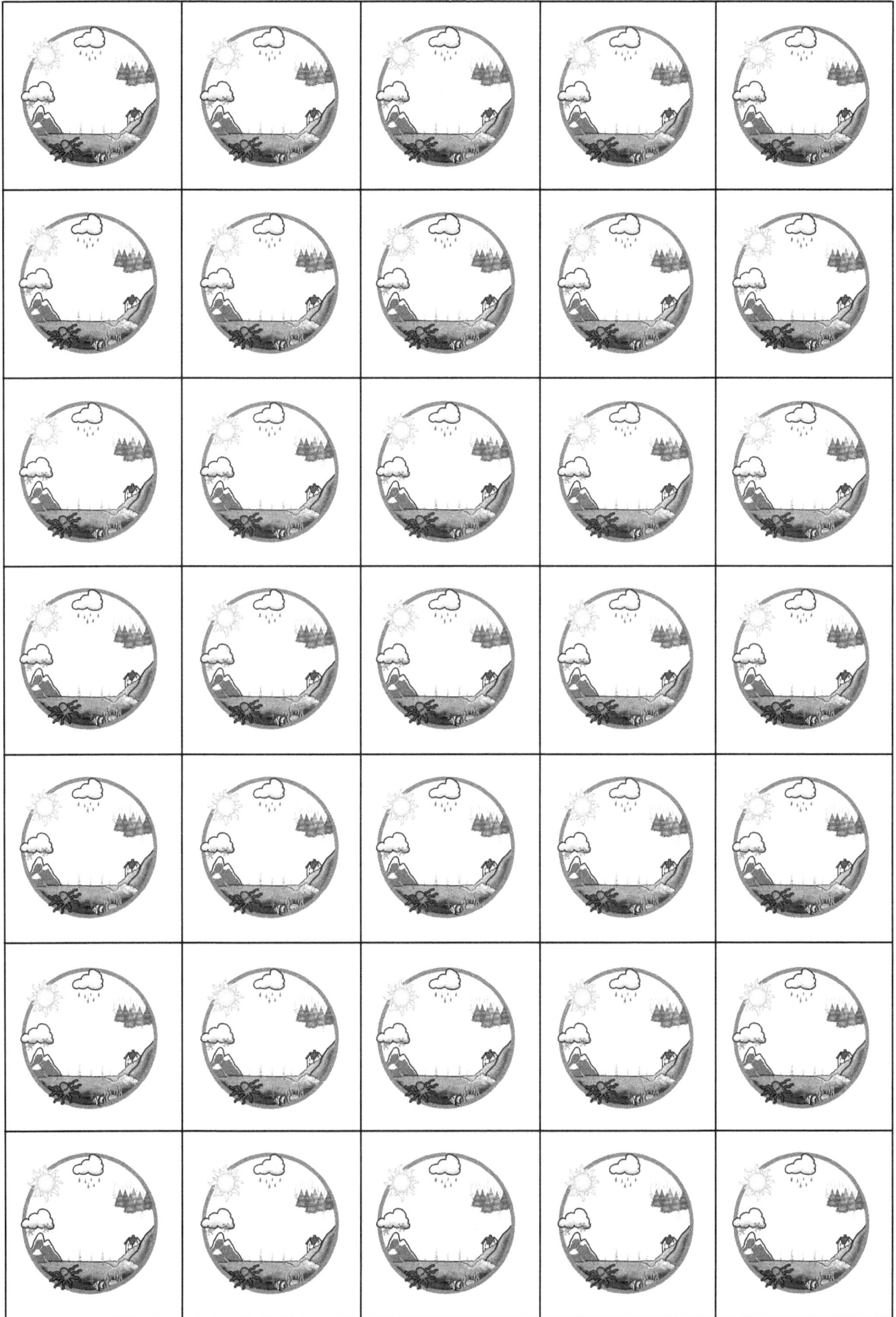

air mass 1. A large body of air that has a similar temperature and about the same amount of moisture is an ___. 2. The boundary between a cold ___ and a warm ___ is called a front.	**air pressure** 1. A barometer is used to measure this. 2. High ___ usually brings good weather.
blizzard 1. This is the name for a severe winter storm. 2. A ___ has strong winds, low temperatures and heavy, blowing snow.	**breeze** 1. A wind with speeds of 4 to 7 miles per hour called a light ___. 2. A wind with speeds of 8 to 12 miles per hour is called a gentle ___.
calm 1. If there is no wind or wind with a speed of less than 1 mile per hour, we call the condition ___. 2. This is the lowest classification on the Beaufort Scale of wind force.	**Celsius** 1. On this scale water freeze(s) at 0 degrees. 2. On this scale water boils at 100 degrees.
Cirrus 1 ___ clouds are the highest clouds. 2. These high clouds are white and feathery. They usually mean fair weather.	**climate** 1. ___ is the average weather over a period of time. 2. ___ refers to the usual weather conditions of a place.
cloud 1. A ___ is condensed vapor in the atmosphere. 2. A ___ is a visible body of fine water droplets or ice particles in the atmosphere.	**cold** 1. We say something is ___ if it has a low temperature. 2. In a ___ front, the leading edge of colder air replaces warmer air.

condensation 1. This is the process of changing from a gas, or vapor, phase into a liquid phase. 2. When water vapor in the air changes back into liquid, forming clouds, it is called ___.	**cumulus** 1. The puffy clouds that look like cotton are ___ clouds. 2. The base of every ___ cloud is usually flat. They grow upwards and form interesting shapes.
deserts 1. Regions that get less than about 10 inches of precipitation a year are called ___. 2. The Sahara and the Mohave are examples of hot ___; however, a few cold ___ are found in higher latitudes.	**dew** 1. We call water droplets that condense on cool surfaces at night ___. 2. The minute droplets that often condense on morning grass is called ___.
direction(s) 1. The four principal ___ are north, south, east and west. 2. Winds are named by the ___ from which they come. An east wind is a wind from the east.	**downpour** 1. It is what we call a heavy rain. 2. A drenching rain is a ___.
drizzle 1. To rain in fine, mist-like drops is to ___. 2. A very fine, misty rain is called a drizzle.	**drought** 1. An extended period of time without rain is called a ___. 2. A scarcity of rain is called a ___. During these times it is important to conserve water.
equator 1. This is the name for the imaginary line that encircles the earth. 2. Temperatures decrease the further away from the ___ you go.	**evaporation** 1. The process of changing from a liquid state to a gas or vapor is ___. 2. When the sun heats the water in rivers, lakes or the ocean and turns it into vapor or steam it is called ___.

Fahrenheit 1. On this scale water freezes at 32 degrees. 2. On this scale water boils at 212 degrees.	**flurries** 1. Light snow showers are sometimes called ___. 2. Complete this analogy: drizzle : rain :: ___ : snow. ("Drizzle is to rain as blank is to snow.")
fog 1. When droplets of water vapor are suspended near the ground, it is called ___. 2. The difference between a cloud and ___ is that ___ forms close to the ground.	**freeze(s)** 1. To convert from a liquid to a solid by cold is to ___. 2. Water ___ at 0° C and 32° F.
front 1. A boundary between two different air masses is called a ___. 2. When 2 air masses meet but neither is strong enough to replace the other, a stationary ___ results.	**gale** 1. A ___ is a very strong wind but not strong enough to be considered a violent-storm wind. 2. If a wind has a speed of 32 to 63 miles per hour, it is classified as a ___-force wind.
hail 1. ___ is precipitation in the form of small balls. 2. ___ has layers of clear ice and compact snow.	**humidity** 1. Dampness in the air is called ___. 2. The amount of water vapor in the air is ___.
hurricane 1. It is a severe storm with winds of 75 miles per hour and greater. 2. In 2005 ___ Katrina devastated the city of New Orleans.	**latitude** 1. We measure the distance north and south of the equator in degrees of ___. 2. The equator is located at 0° ___.

Weather Bingo

lightning 1. It is an electric discharge in the atmosphere that produces a flash of light. 2. This electric discharge can be from cloud to cloud or cloud to ground.	**meteorologist** 1. A scientist who studies the weather is called a ___. 2. A weather forecaster is a ___.
overcast 1. When the sky is covered with clouds, it is said to be ___. 2. Individual clouds are not usually seen in an ___ sky.	**precipitation** 1. Rain, snow, sleet and hail are forms of ___. 2. ___ occurs when so much water has condensed that the air cannot hold it anymore.
rain 1. Precipitation is called ___ when it falls as drops of water if the drops are .5 mm or more. 2. Light ___ or snow is called a shower.	**rainbow** 1. It is an arc of colored light. 2. The colors of the ___ are red, orange, yellow, green, blue, indigo and violet.
revolution 1. The time it takes for Earth to make one ___ around the sun is a year. 2. The ___ of Earth around the sun as it rotates on a tilted axis causes the seasons.	**seasons** 1. The ___ are summer, autumn, winter and spring. 2. Summer is the warmest ___; winter is the coldest.
snow 1. ___ is precipitation in the form of ice crystals. 2. ___flakes are 6-sided ice crystals.	**storm(s)** 1. A disturbance in the atmosphere that includes strong winds and heavy precipitation is a ___. Some also include thunder and lightning. 2. Hurricanes, blizzards, ice ___ and thunder___ are some examples.

Weather Bingo

stratus 1. ___ clouds are low-altitude clouds. They appear as a horizontal layer of grey. 2. ___ are large dark clouds that are low in the sky.	**sun** 1. Earth and the other planets revolve around it. 2. We get our heat and light from the ___.
Temperate 1. The continental United States is in the North ___ Zone. 2. The ___ Zone features cold winters and mild summers.	**temperature** 1. It is the degree of hotness or coldness of a body or an environment. 2. We use a thermometer to measure ___.
thunder 1. We hear the ___ before we see the lightning. 2. It is the loud sound produced by expanding air along the path of lightning.	**tornado** 1. A ___ is a violently rotating column of air. Some have wind speeds over 200 mph. 2. A ___ is characterized by a funnel-shaped cloud. It is very dangerous.
Tropics 1. This region is also called the Torrid Zone. 2. It is hot and wet in this region, which includes the equator.	**water cycle** 1. The circulation of water on, above and below Earth is called the ___. 2. Evaporation, condensation, precipitation and collection (in the ground and in oceans, lakes and rivers) are processes in the ___.
weather 1. It is the state of the atmosphere at a given time and place. 2. ___ is the state of the atmosphere with respect to temperature, mois-ture, wind, degree of cloudiness and other factors. Weather Bingo	**wind** 1. ___ is moving air. 2. A breeze is a light ___. A gale-force ___ is strong. © Barbara M. Peller

Weather Bingo

equator	evaporation	flurries	water cycle	thunder
climate	air pressure	Tropics	humidity	Fahrenheit
storm(s)(s)	precipitation		fog	latitude
wind	air mass	drizzle	temperature	freeze(s)
front	weather	cold	downpour	drought

Weather Bingo: Card No. 1

© Barbara M. Peller

Weather Bingo

temperature	sun	gale	overcast	front
freeze(s)	humidity	Cirrus	air mass	snow
rainbow	weather		condensation	drizzle
direction(s)	revolution	precipitation	lightning	Fahrenheit
drought	Tropics	cold	climate	downpour

Weather Bingo: Card No. 2

Weather Bingo

wind	drizzle	humidity	temperature	storm(s)
weather	air pressure	calm	evaporation	hurricane
air mass	Tropics		snow	blizzard
precipitation	rainbow	front	direction(s)	gale
downpour	climate	cold	lightning	flurries

Weather Bingo

precipitation	snow	weather	climate	flurries
hail	Cirrus	evaporation	overcast	storm(s)
fog	direction(s)		thunder	temperature
drizzle	dew	Tropics	cold	calm
deserts	drought	meteorologist	downpour	latitude

Weather Bingo

drought	thunder	air mass	Cirrus	climate
hail	drizzle	calm	condensation	air pressure
sun	latitude		cumulus	flurries
Fahrenheit	snow	equator	lightning	deserts
humidity	cold	rain	precipitation	fog

Weather Bingo

blizzard	snow	gale	sun	latitude
temperature	air mass	deserts	evaporation	storm(s)
overcast	calm		Cirrus	condensation
cold	front	lightning	meteorologist	fog
freeze(s)	drizzle	equator	rain	flurries

Weather Bingo

equator	snow	temperature	cumulus	humidity
freeze(s)	flurries	weather	air pressure	hail
gale	wind		condensation	breeze
precipitation	direction(s)	storm(s)	stratus	rainbow
cold	climate	lightning	meteorologist	blizzard

Weather Bingo

fog	snow	Celsius	temperature	breeze
hail	sun	overcast	latitude	Cirrus
storm(s)	seasons		flurries	thunder
downpour	precipitation	weather	deserts	direction(s)
Tropics	cold	meteorologist	air mass	freeze(s)

Weather Bingo: Card No. 8

Weather Bingo

condensation	humidity	weather	storm(s)	latitude
deserts	sun	fog	air mass	flurries
hurricane	equator		air pressure	Celsius
breeze	drought	front	cumulus	stratus
direction(s)	lightning	calm	temperature	thunder

Weather Bingo

stratus	water cycle	Cirrus	overcast	rain
latitude	breeze	evaporation	air pressure	flurries
seasons	snow		wind	rainbow
front	Fahrenheit	deserts	lightning	hurricane
cloud	freeze(s)	gale	drought	fog

Weather Bingo

blizzard	snow	air mass	deserts	freeze(s)
Celsius	hurricane	cumulus	condensation	evaporation
hail	sun		gale	weather
cloud	storm(s)	lightning	climate	temperature
calm	cold	equator	meteorologist	humidity

Weather Bingo

humidity	thunder	hurricane	stratus	condensation
weather	Tropics	sun	meteorologist	air pressure
equator	temperature		latitude	overcast
cold	direction(s)	flurries	wind	hail
snow	Celsius	seasons	calm	breeze

Weather Bingo

cloud	thunder	blizzard	hurricane	latitude
sun	Celsius	snow	condensation	rainbow
wind	Cirrus		weather	temperature
fog	lightning	breeze	seasons	stratus
cold	Fahrenheit	meteorologist	equator	cumulus

Weather Bingo

climate	sun	air mass	condensation	cloud
breeze	equator	hurricane	air pressure	snow
deserts	temperature		gale	calm
Fahrenheit	lightning	seasons	Cirrus	blizzard
cold	overcast	rainbow	freeze(s)	fog

Weather Bingo

cumulus	condensation	air mass	humidity	wind
blizzard	gale	evaporation	sun	deserts
latitude	equator		storm(s)	flurries
cold	hurricane	Celsius	lightning	cloud
freeze(s)	direction(s)	meteorologist	rain	weather

Weather Bingo

Cirrus	hurricane	Celsius	rain	revolution
overcast	rainbow	stratus	hail	water cycle
cloud	thunder		latitude	weather
precipitation	breeze	cold	cumulus	wind
deserts	tornado	meteorologist	direction(s)	snow

Weather Bingo: Card No. 16

Weather Bingo

cloud	Temperate	dew	hurricane	climate
cumulus	deserts	lightning	stratus	temperature
condensation	wind		tornado	Celsius
drought	freeze(s)	fog	air mass	rainbow
front	calm	humidity	water cycle	thunder

Weather Bingo

flurries	seasons	breeze	deserts	overcast
snow	cloud	front	latitude	calm
condensation	rainbow		dew	rain
drought	evaporation	lightning	wind	gale
tornado	hurricane	air mass	Temperate	blizzard

Weather Bingo

latitude	blizzard	hurricane	Celsius	seasons
cumulus	wind	rain	humidity	water cycle
Temperate	climate		air pressure	drought
gale	tornado	front	direction(s)	dew
storm(s)	revolution	freeze(s)	fog	meteorologist

Weather Bingo

seasons	Temperate	water cycle	hurricane	air pressure
Cirrus	weather	hail	front	overcast
thunder	stratus		precipitation	evaporation
drought	fog	downpour	direction(s)	tornado
drizzle	Tropics	revolution	wind	dew

Weather Bingo

cumulus	blizzard	hail	hurricane	Fahrenheit
thunder	dew	breeze	Celsius	equator
rainbow	freeze(s)		Temperate	air mass
front	humidity	tornado	drought	fog
precipitation	revolution	meteorologist	cloud	direction(s)

Weather Bingo

storm(s)	gale	dew	sun	cloud
overcast	water cycle	flurries	Celsius	air pressure
breeze	wind		equator	stratus
tornado	drought	direction(s)	evaporation	climate
revolution	calm	Temperate	rainbow	hail

Weather Bingo: Card No. 22

Weather Bingo

Cirrus	Temperate	humidity	sun	meteorologist
blizzard	seasons	freeze(s)	cumulus	evaporation
gale	cloud		downpour	equator
rainbow	revolution	tornado	calm	direction(s)
Fahrenheit	fog	Tropics	front	dew

Weather Bingo

Cirrus	seasons	climate	Temperate	Celsius
latitude	meteorologist	hail	overcast	equator
stratus	rain		cloud	rainbow
Fahrenheit	downpour	tornado	calm	thunder
drizzle	precipitation	revolution	water cycle	Tropics

Weather Bingo

precipitation	hail	Temperate	air mass	dew
evaporation	Fahrenheit	cumulus	Cirrus	air pressure
thunder	Celsius		downpour	tornado
rain	drought	Tropics	revolution	water cycle
meteorologist	climate	breeze	deserts	drizzle

Weather Bingo: Card No. 25

Weather Bingo

dew	Temperate	downpour	overcast	rain
front	wind	Celsius	seasons	Cirrus
Fahrenheit	gale		water cycle	precipitation
cloud	sun	drought	revolution	tornado
stratus	deserts	air mass	Tropics	drizzle

Weather Bingo

downpour	breeze	Temperate	seasons	weather
Fahrenheit	gale	cumulus	tornado	air pressure
lightning	Tropics		revolution	precipitation
rain	blizzard	hail	drizzle	evaporation
cloud	water cycle	dew	storm(s)	stratus

Weather Bingo

latitude	seasons	temperature	Temperate	breeze
weather	dew	downpour	front	water cycle
Tropics	rainbow		rain	overcast
stratus	storm(s)	freeze(s)	revolution	tornado
sun	condensation	cloud	drizzle	Fahrenheit

Weather Bingo

dew	seasons	rain	cumulus	condensation
Fahrenheit	front	hail	stratus	storm(s)
thunder	downpour		air pressure	Temperate
weather	drought	flurries	revolution	tornado
Cirrus	Celsius	drizzle	blizzard	Tropics

Weather Bingo: Card No. 29

Weather Bingo

climate	Temperate	overcast	condensation	tornado
evaporation	rain	gale	water cycle	air pressure
drizzle	calm		temperature	hail
Fahrenheit	blizzard	seasons	revolution	downpour
drought	humidity	Tropics	dew	flurries

Weather Bingo: Card No. 30

www.ingramcontent.com/pod-product-compliance
Lightning Source LLC
Chambersburg PA
CBHW051420200326
41520CB00023B/7308